Measurements in Physics

© sciencebod 2016

Preface

Almost everything we do in Physics is about measurements. In this book, we present very fundamental introductions to physical quantities.

The distinction between fundamental and derived quantities is well explained using very numerous examples. There is very excellent tutorial on how to take measurements with the vernier caliper and with the micrometer screw gauge. The book also presents the ideas of units and dimensions in a manner that is useful even for students taking advanced courses in Physics.

Our thinking is that it is very important for science students to understand fundamental ideas about measurements and to be able to use this understanding to drive their learning in different other aspects of science. The book presents very clear explanations of things we think students ought to know about measurements of physical quantities. There are also numerous examples to guide the students' learning, and exercises to test their understanding of the topic.

© sciencebod 2016; okodan2003@gmail.com; +2348136094616, +2348062111865

Measurements

1.1 Quantities and their units

> **1**

Nearly everything we do in Physics is about measurements. It's either we are measuring the distance between two places (length), the quantity of matter contained in an object (mass), the time interval between two events (time), the ability to do work (energy), the amount of space in a container (volume), etc.

These physical quantities are grouped into two, according to whether they are directly measurable or dependent on other ones: these are fundamental and derived quantities.

Fundamental quantities

> **2**

Fundamental quantities are those that can be measured directly, and whose measurements do not depend on any other quantity. Examples include:

Length (which can be measured directly using a rule)
Mass (which can be measured directly using a beam balance)
Time (which can be measured directly using a stop watch)

Hey! Can you measure volume directly?
The answer is No. Why?

Derived quantities

> **3**

Yes! We cannot measure volume directly because it is not a fundamental quantity. Let's explain more: in order to measure the volume of a rectangular block for example, we measure the length, width and breadth of the block first.

Then we multiply the length by the width and the breadth to get the volume of the block.

Note: Length, width and breadth are all length which is a fundamental quantity. Therefore, volume is derived from length.

Derived quantities are those quantities whose measurements depend on those of other quantities. Examples of derived quantities are: Area, Volume, Density, Velocity, Acceleration, etc.

Do quantities have units

4

The answer is yes.
Units are used to clear the ambiguities involved in describing quantities. For example, if asked how much water you have in the house and you replied 500, I'll go like 500 what? Litres? Kilograms? Gallons? ...

Units give someone the idea of exactly what is being talked about when dealing with quantities, and exactly how much of the quantity is referred. Units add meanings to numbers or quantities.

S.I. Units

5

The system of units accepted internationally is known as the System International (S. I.) units.
The S. I. unit of mass is kilogram (kg), for length is meters (m) and that of time is second (s).

Do units only exist in S. I. form?

6

No! There are multiples and sub multiples of the S. I. units. For example; kilometer (1000 metres), hour (60×60 seconds), millimeter (0.001 metres), gramme (0.001kg), millisecond (0.001s) etc.

In a nutshell

7

We can sum up the fundamental and derived quantities, their units and symbols in the following tables.

Table 1. Fundamental quantity, symbol and unit.

Quantity	Unit	Symbol
Length	Meter	m
Mass	Kilogram	kg
Time	Second	s
Temperature	Kelvin	K
Electric current	Ampere	A
Amount of substance	Mole	Mol

Table 2. Derived quantity, derivation and unit.

Quantity	Derivation	Unit (Symbol)
Area	Length × breadth	m^2
Volume	Length × breadth × height	m^3
Density	Mass/volume	kg/m^3
Speed or velocity	Distance/time	m/s
Acceleration	Change in velocity/time	m/s^2
Force	Mass × acceleration	kgm/s^2 = Newton (N)
Momentum	Mass × velocity	kgm/s = Newton second (Ns)
Impulse	Force × time	Ns
Pressure	Force/area	N/m^2 = Pascal (Pa)
Energy or work	Force × distance	Nm = Joules (J)
Power	Work/time	Nm/s = Watts (W)
Heat capacity	Quantity of heat/temperature	J/K
Linear expansivity	Increase in length / (original length × temperature difference)	m/mK = K^{-1}

Note!

| 8 |

Plan 7 illustrates just a few of the most commonly used fundamental and derived units, their symbols and units.

1.2 Measuring Instruments

| 9 |

Here we are going to look into the instruments used to measure the fundamental quantities namely: length, mass and time.

Hey! Why not derived quantities?
Oh! Derived quantities are not directly measurable, though advances in technology today has made it possible to produce instruments that can be used to measure derived quantities.

1.2.1 Measurement of Length

| 10 |

Most often, length is measured with the following instruments:
1. Rule (also called 'Ruler')
2. Vernier caliper
3. Micrometer screw gauge

The suitability of any of these instruments is determined by the size of what is being measured. We will then look at them one after the other.

The rule

| 11 |

Long distances are measured with tapes and rules that are graduated up to one meter (meter rule), i.e. 100cm equivalent.

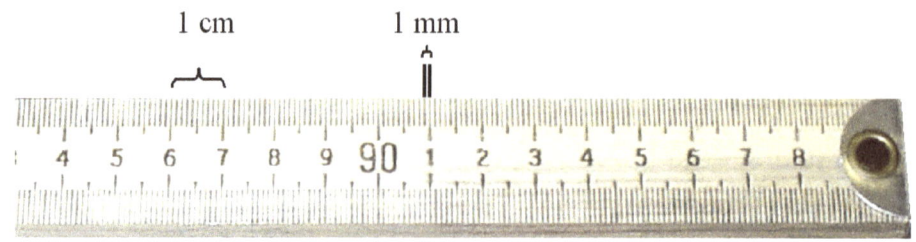

Fig 1. Section of a meter rule, showing its graduations

Meter rules are graduated in centimeters and millimeters.
Note: The reading accuracy of a measuring instrument is found by dividing the least graduation of the instrument by 2.

Can we then find the reading accuracy of a meter rule? Here it is: since the least graduation of a meter rule is 1mm then its reading accuracy is ½ mm = 0.5mm. Yes! That's how it works.

Question (JAMB)

12

What is the least possible error in using a rule graduated in centimeters?
(A) 0.1cm, (B) 0.5cm (C) 1.0cm (D) 2.0cm

Answer: Option (B) is correct! 0.5cm
These are the steps: first we are only told that the rule is graduated in centimeters, so the least graduation on the rule is 1 cm, then dividing by 2 gives ½ cm = 0.5cm.
Great!

The Vernier calipers

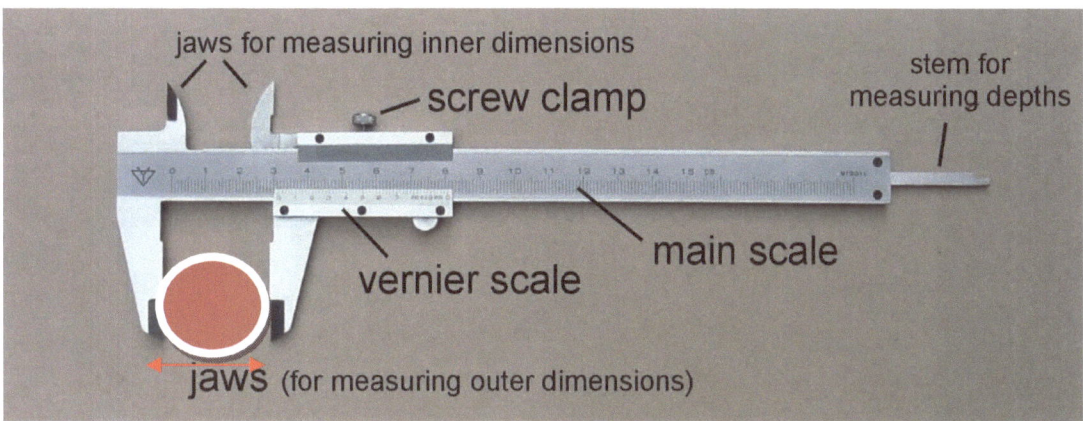

Fig 2. The Vernier Calipers

Vernier Calipers measure more accurately than the metre rule. Their accuracy is 0.01cm, meaning they can tell you the difference between two lengths that are; say 2.41cm and 2.42cm. A meter rule will only be able to tell you that both lengths are 2.4cm.

One key importance of the vernier calliper is that it can be used to measure both internal and external dimension of objects, as well as depths.

Structure and components

Vernier calipers have two scales, the main scale, M and the vernier scale, V.

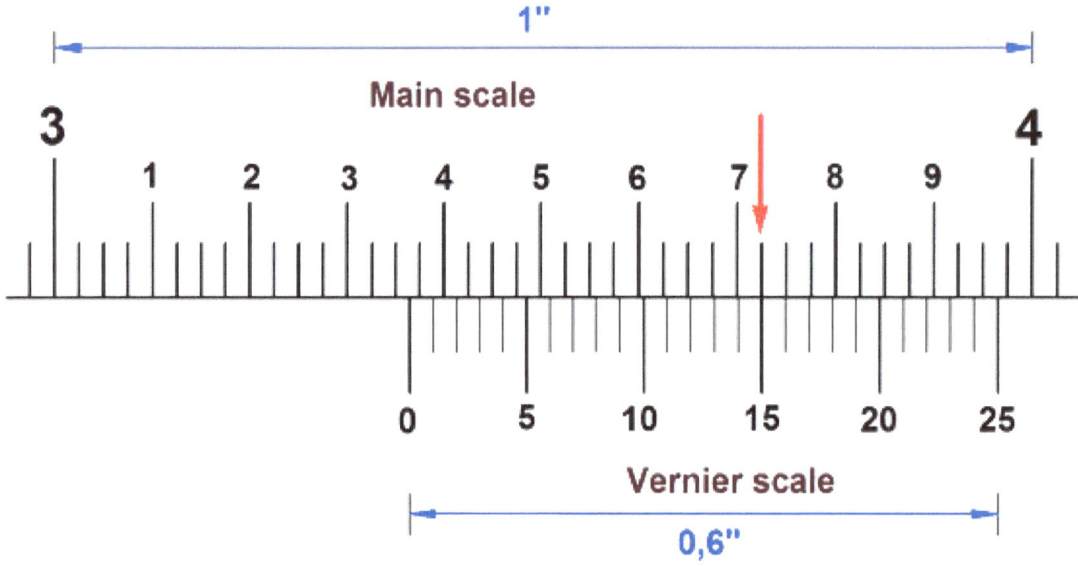

Fig 3. Vernier Caliper with Vernier and Main Scale

How do we use the vernier caliper?

The next plan has the answer.

Plan 15!

15

The instrument is used in the following ways:

To measure the external dimensions of an object, the object is placed between the lower jaws, as shown in figure 2 which are then moved together until they hold the object.

To measure the internal dimensions, the object is placed over the upper jaws, which are then moved apart until they hold the object.

And to measure a depth, the stem of the caliper is put into the depth and its length adjusted until it exactly fills the depth.

How is it then read? Get the steps!

16

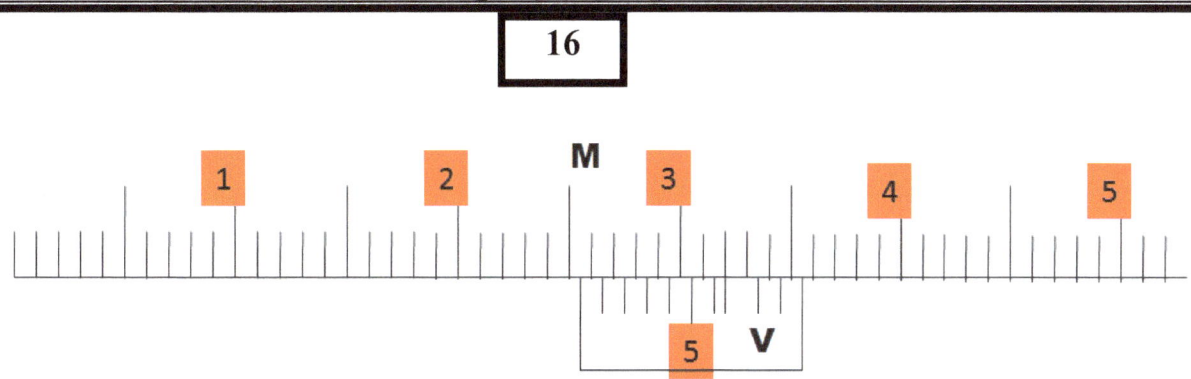

To take the vernier caliper reading from the figure above, we do the following:

1. First note the main scale reading just before the vernier scale; that is 2.5cm.

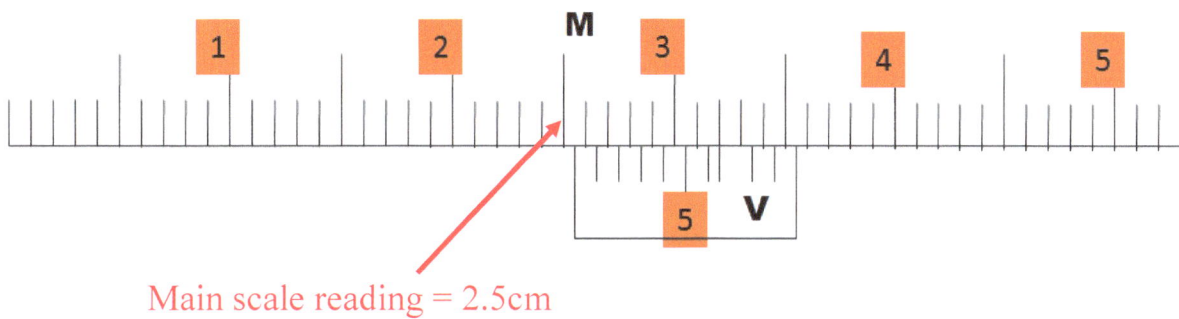

Main scale reading = 2.5cm

2. Then note the vernier scale reading that aligns with a reading on the main scale; that is 0.07cm

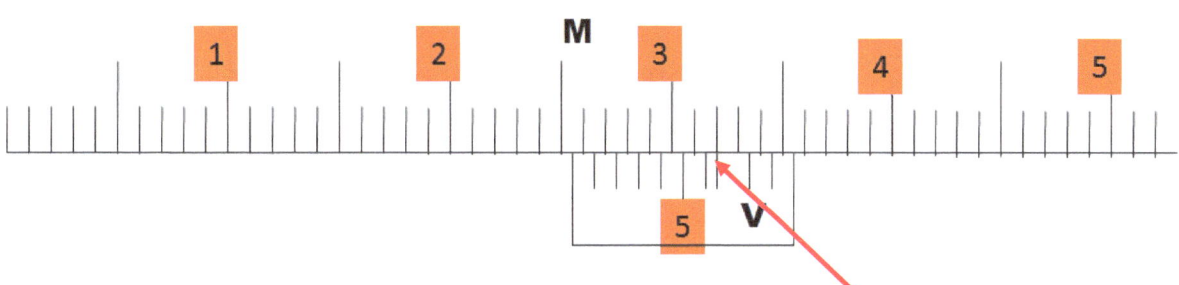

The 7th mark aligns with a mark on the main scale, therefore the vernier scale reading = 0.07cm

Note that all other marks on the vernier scale DO NOT align with any of the marks on the main scale.

3. Add up the main scale and vernier scale readings;
that is, 2.5cm + 0.07cm = 2.57cm

Yes! That is the overall reading of the vernier caliper, and that is how to take vernier caliper readings.

Great if you got that! Let's take some more examples.

Next Example!

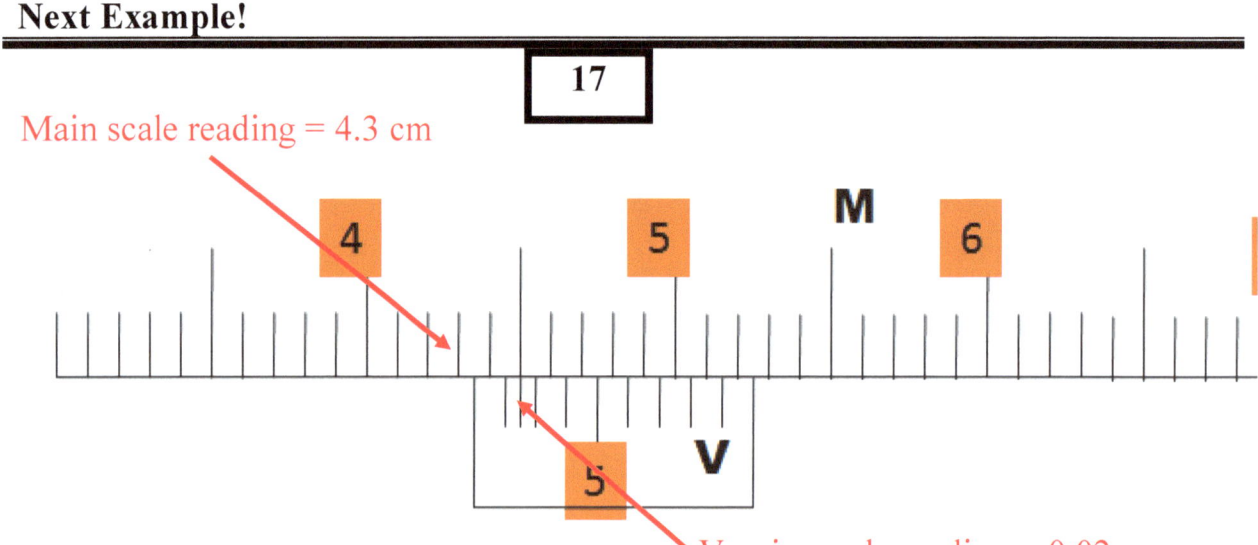

First, the main scale reading 4.3 cm, and the vernier scale reading is 0.02 cm as illustrated in the diagram above. Therefore the overall vernier caliper reading is:

4.3 cm + 0.02 cm = 4.32 cm.

Great if you understood that!
Now, try the next one all on your own!

What's the reading of this vernier caliper?

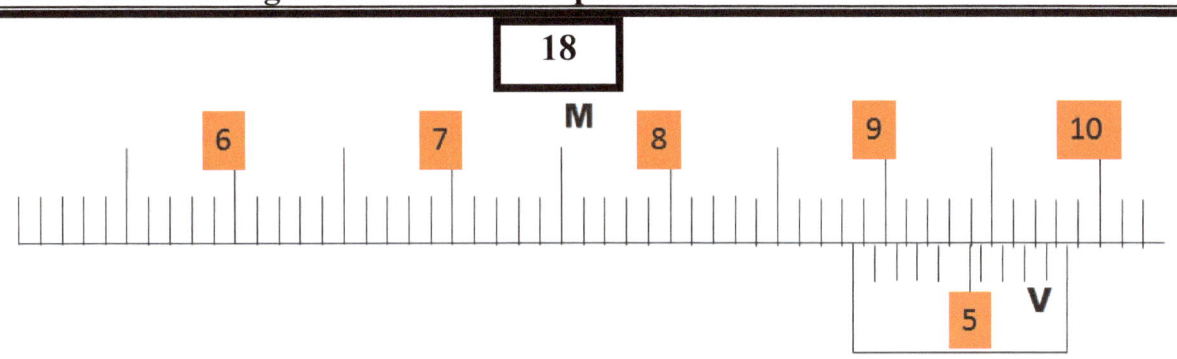

Your answer?
See if you got it in the next plan!

Did you get it correctly?

Yes! You did if you got 8.85 cm.

The main scale reading was 8.8 cm, while the vernier scale reading was 0.05 cm.

Congratulations if you got that!

You could try to take more vernier caliper readings at the Exercise section. Meanwhile, let's proceed with the micrometer screw gauge.

The micrometer screw gauge

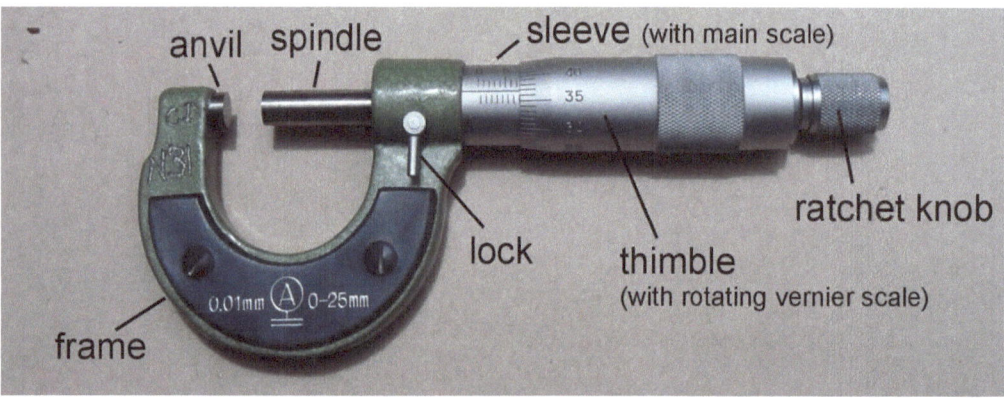

Fig 4. The micrometer screw gauge

The micrometer screw gauge, measures even smaller lengths than the vernier caliper. Its reading accuracy is higher than that of the vernier caliper.

It can measure to an accuracy of 0.001cm.

This makes it suitable for measuring the thickness of very tiny objects like the thickness of a thin wire, thickness of a sheet of paper, etc.

Structure and components in brief

The micrometer screw guage has an anvil and spindle which functions the same way as the jaws of a plier.

Like the vernier caliper, the micrometer screw guage has a main and a vernier scale. The micrometer screw gauge is usually graduated in millimeters because of the small lengths it is often used to measure.

It also has a ratchet which is used to hold objects firmly between the jaws (anvil and spindle).

Now you are ready to learn how to read the instrument.

Here we go

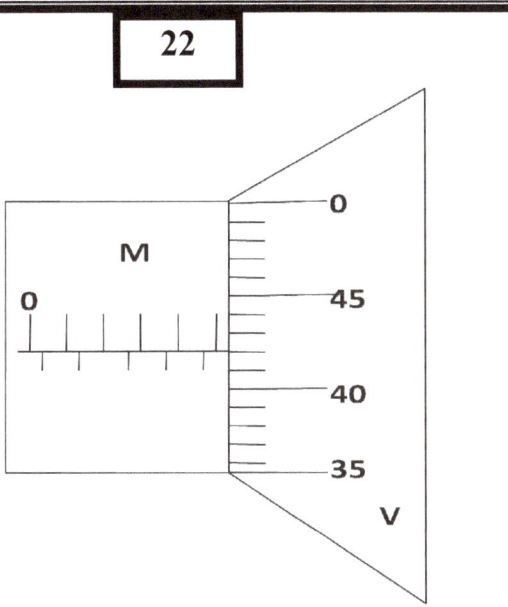

The main scale is marked M while the vernier scale is marked V.

1. First, we have to read the main scale:

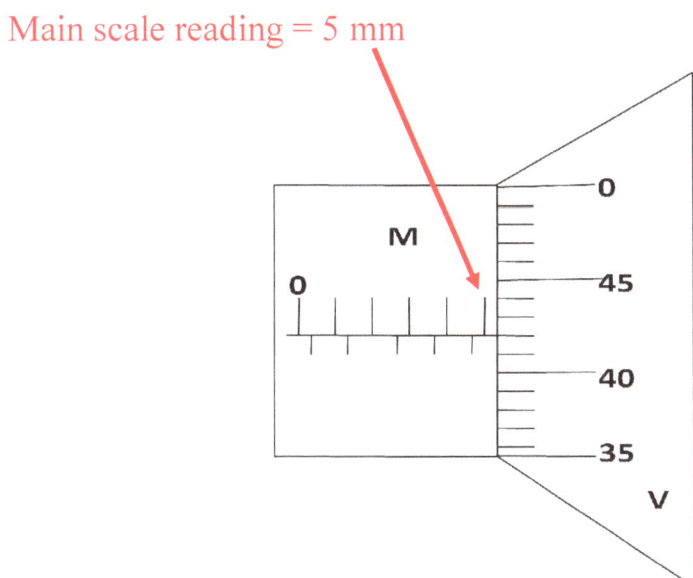

Main scale reading = 5 mm

2. Next, we read the vernier scale (that is the mark that aligns with the horizontal line on the main scale).

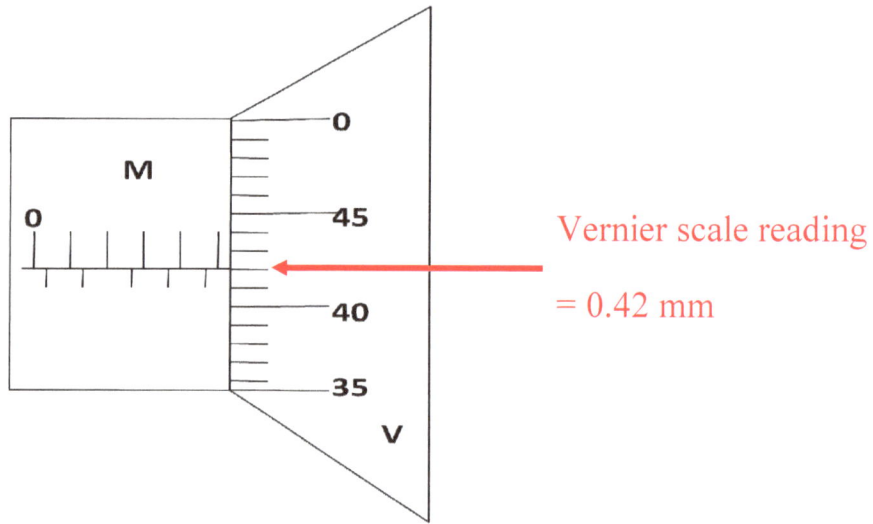

3. And finally, the overall micrometer screw gauge reading is the sum of the main and vernier scale readings

= 5 mm + 0.42 mm

= 5.42 mm

Fantastic!

Note that 5.42 mm is the same as 0.542 cm. So as we said earlier, the micrometer screw gauge can read to an accuracy of 3 decimal centimeters (That is, 0.001 cm).

Next try!

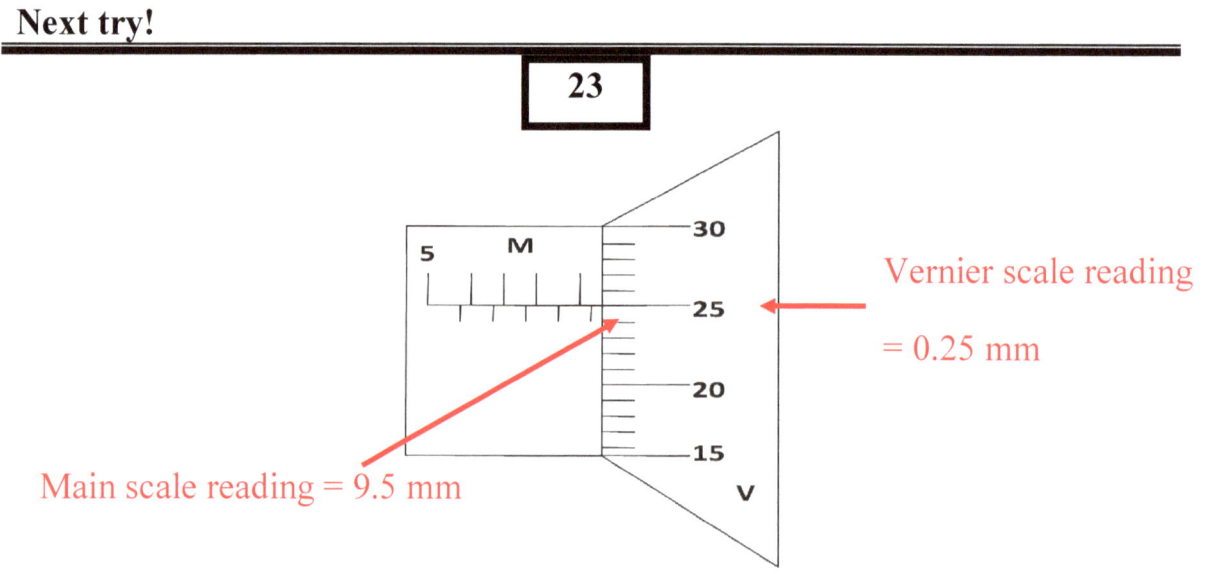

For the micrometer screw gauge above, the main scale reading is 9.5 mm and the vernier scale reading is 0.25 mm, therefore the overall micrometer screw gauge reading is
9.5 mm + 0.25 mm
=9.75 mm.

Next one for you to try all by yourself!

For you to try!

24

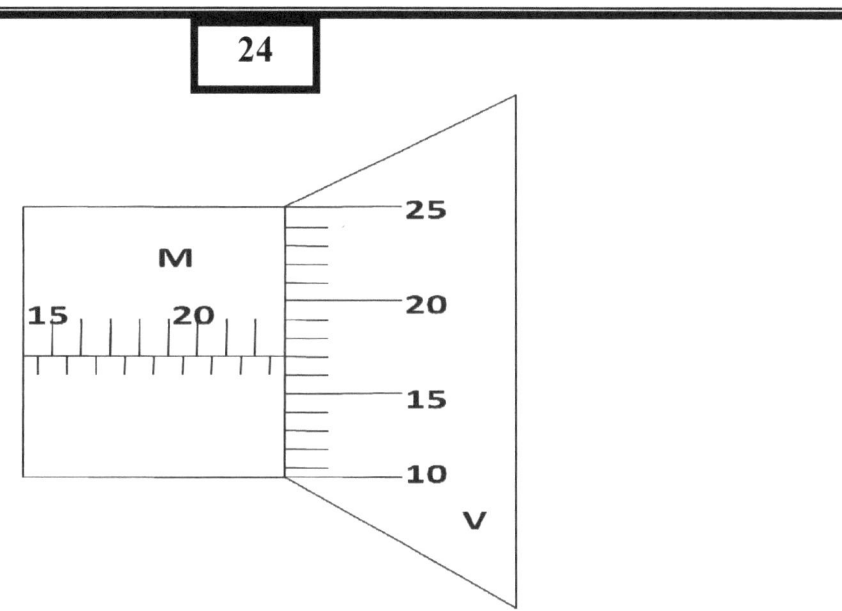

If you got the answer in the next plan, then you're good to go!

Answer!

25

The main scale reading is 22.5 mm, while the vernier scale reading is 0.17 mm.

Therefore, the overall micrometer screw gauge reading is
= 22.5 mm + 0.17 mm
= 22.67 mm

Great if you got that!
You can practice more of the micrometer screw gauge readings at the Exercise

1.2.2 Measurement of Mass

section.

The mass of an object can be measured with a beam balance. The beam balance has two scale pans, X and Y.

Fig 5. The Beam Balance

To use the beam balance, the object whose mass is to be determined is placed on the scale pan (X) of the beam balance, and known masses are added into the other pan (Y) until the beam is horizontal and does not fall to the side of either scale pan. The total mass in Y is then equal to the value of the unknown mass in X.

The physical principle involved in the working of the balance is the principle of moments.

Reading accuracy of the beam balance

The reading accuracy of the beam balance could be up to 0.001grams depending on the sensitivity. This sensitivity depends on the smallest values of the known masses in pan Y.
[Just for your information; the known masses are also referred to as standard masses].

If we use smaller values of the known masses, then we can get the unknown mass of the object in pan X to a better accuracy. Let's say we use known masses with values as low as 0.001grams, then we can get the value of the unknown mass to that accuracy. This in return represents the reading accuracy of the beam balance.

The Spring Balance

28

The spring balance is originally designed to measure the weight of an object, but the weight of an object and its mass are directly related for any given location on Earth. [This has been well explained in our book on 'Newton's Laws of Motion']. So, we can also use the spring balance to measure the mass of an object.

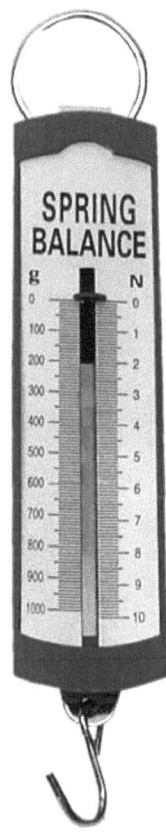

Fig 6. The Spring Balance

The spring balance mainly comprises of a spring to which a hook is attached at the bottom.

The principle is basically that heavier objects will cause more extensions on the spring; this is from Hooke's law, which has been well treated in our book on 'Elasticity'.

To measure the mass of an object, the object is simply hung on the hook and its mass is read from the scale.

Designers of most spring balances assume the value of the acceleration due to gravity to be about $10 m/s^2$, so with this value, they correlate the weights of objects with their masses. This is illustrated on the scale of the spring balance in Figure 6; we have weights (in N) on one side, and masses (in g) on the other side.

Now, let's try our hands on the following question

29

Which of the following statements about a spring balance and/or a chemical balance is not correct?
(a) The chemical balance operates on the principle of moments.
(b) The spring balance operates on Hooke's law
(c) Either may be used to measure weight of a substance.
(d) The reading of a spring balance changes with altitude.

The answer is C:
The chemical balance is only used to measure mass and not weight. Unless we also measure the value of acceleration due to gravity, then we can multiply it with the mass to get weight.

Yes, the reading of a spring balance changes with altitude. This is because the weight of an object will change with altitude since the value of acceleration due to gravity also does.

1.2.3 Measurement of Time

30

All events which happen in nature involve the idea of time. We can define time as that in which events are distinguishable with reference to before and after. The most natural time unit is the solar day which is manifested by the passing of day and night.

In the laboratory, time is measured with a stop – clock or stop watch.

Fig 7. The Stop Watch

How does the stop watch or stop clock work?

The stop watch

31

The stop watch is started or stopped by pressing one of the knobs (different manufacturers have different options of which of the knobs to press).

The stop watch enables us to measure small intervals of time very accurately. Typical stop watches have reading accuracies of 0.1 seconds. Some recent electronic stop watches have reading accuracies as small as 0.001 seconds, or even smaller.

The unit of time as we already know is the second (s), with other multiple units as:
60 seconds = 1 minute
60 minutes = 1 hour
24 hours = 1 day, etc.

1.3 Measurement of derived quantities

32

As we have said earlier, derived quantities are measured by a combination of fundamental quantities using formula.
Here we want to see how derived quantities (like area and volume) are measured.

1.3.1 Measurement of area

33

Areas of regular figures are determined by measuring lengths in two dimensions and using the formula.
The formula for the areas of some regular figures are
- (i) Triangle = ½ × base × height,
- (ii) Rectangle = length × breadth,
- (iii) Parallelogram = height × length of one parallel sides.
- (iv) Trapezium = ½ (sum of the length of the parallel sides) × height
- (v) Circle = π × radius^2

The unit of area is metre^2 (m^2)

1.3.2 Measurement of volume

34

Volume is obtained by measuring length in three dimensions, and using appropriate formula as follows:
- (i) Rectangular box = length × breadth × thickness
- (ii) Cylinder = π × radius² × height
- (iii) Cone = ⅓ × π × radius² × height
- (iv) Sphere = $\frac{4}{3}$ × π × radius³

And the unit of volume is metre³ (m³)

Volume of liquids

35

The volumes of quantities of liquids can be determined by using measuring cylinders.

Fig 8. The Measuring Cylinder

The measuring cylinder is usually calibrated in cubic centimeters cm³, or milliliters, ml, which are equivalent units.

Can we measure the volume of irregular bodies?

36

The volume of irregular solids can be obtained by completely immersing the solid in a cylinder containing a liquid which does not dissolve the solid. The volume of the liquid displaced represents the volume of the solid.

Fig 9. Measuring the volume of an irregular object

In the first measuring cylinder, the object has not been immersed, and the volume of liquid in it is 4.8cm³. When the object is immersed as shown in the second measuring cylinder, the volume rises up to 5.6cm³. This means that the volume of the irregular solid is 5.6cm³ – 4.8cm³ = **0.8cm³**.

The volume of the liquid displaced is equal to the volume of the object.

1.4 Measurement errors

37

Can we have measurements that are completely free of errors?

No! Nature has made it that we cannot get perfect measurements that are completely free of errors. A measurement is therefore, only accurate up to a certain degree depending on the instrument used and the physical constraints of the observer. No measurement in physics is complete, or of real practical value unless an indication is given of the range of values within which the true value may lie.

Experimental errors arise from three major sources: systemic, random and erratic sources.

Now we go ahead to explain each of them briefly.

1.4.1 Systemic errors

38

Systemic errors are uncertainties in the measurements of physical quantities due to instruments, or faults in the surrounding conditions. An example of systematic error is the zero-error.

This is the type of error that results when the measuring device gives a value (that is not zero) for an empty quantity, e.g. an ammeter reads 0.1A when it is not connected to a circuit and no current flows through it. Another example is that the start of a meter rule is the 0.2cm mark.

If this error is not taken care of before using the device to take measurements, it affects all the measurements in like manner. A zero error of 0.5mm on a meter rule, for example, makes all measurements with the rule to be 0.5mm more than their actual values.

If we take note of the zero-error before using the device, we can always correct for the error by subtracting the value from all measurements taken with the device.

1.4.2 Random errors

39

Random errors are uncertainties in a measurement due to in-homogeneity in the observed object. For example, the diameter of a long construction wire measured six times at different places along the length may not be the same due to in-homogeneity.
The reading obtained may sometimes be greater than the actual value and at other times smaller than the actual value. (That is, it could be positive or negative).

1.4.3 Erratic errors

40

Erratic errors also affect experimental observations in irregular ways, but often arise from mistakes, unlike random errors which arise from irregular sources other than mistakes.
For example, a student who intends to press the multiplication key of a calculator may press the division key by mistake and this will constitute erratic error to the measured value.

Next! How do we assess experiment errors?

1.4.4 Assessment of Experimental errors

41

It is necessary to indicate how precise a measurement is and this is always shown by the number of decimal places to which the value is given. For example, if a metal rod is measured using a meter rule and its length recorded as 26.1cm, it means that the length is only accurate to 0.1cm. To indicate this, the value may be written as (26.1 ± 0.1) cm.

Similarly, a measurement of 44.62mm obtained using a micrometer screw gauge is written as (44.62 ± 0.01) mm to denote uncertainty or error of 0.01mm.

Fractional and Percentage Errors

42

Errors can also be stated as fractions of the measured values. This is called fractional error. Fractional errors are calculated as $\frac{error\ value}{measured\ value}$

For example, in the measurement of (44.62 ± 0.01) mm, the fractional error is $\frac{0.01}{44.62} = 2.24 \times 10^{-4}$.

Similarly, percentage errors are defined as $\frac{error\ value}{measured\ value} \times 100$

And in the example above, the percentage error is therefore calculated as: $\frac{0.01}{44.62} \times 100 = 0.0224\%$.

Now, let's take this JAMB question

43

The external and internal diameters of a tube are measured as (32 ±2) mm and (21±1) mm respectively. Determine the percentage error in the thickness of the tube.
(a) 27%   (b) 14%   (c) 9%   (d) 3%

Solution:
The thickness of the tube is = external diameter – internal diameter
= 32mm – 21mm = 11mm

The error in the thickness is (2 - 1) mm = 1mm

Therefore, the thickness with its error is written as: (11 ± 1) mm

And the percentage error in the thickness is therefore $\frac{1}{11} \times 100\% = 9\%$

So, option C is correct.

1.5 Dimensions of physical quantities

44

The dimension of a physical quantity shows how the physical quantity is related to the fundamental quantities namely: mass (M), length (L), and time (T).

For example, the dimension of Area is obtained as below:
Area = length (L) × width (L) = L × L = L^2

Next! In what ways can we utilize the knowledge of dimensions?

1.5.1 Applications of Dimensions

45

Three major applications of the knowledge of dimensions are:
  (i)   To verify the correctness of a physical equation (physical equations are dimensionally homogeneous).
  (ii) To derive the unit of a physical quantity.
  (iii) To derive the exact form of a relationship between measured physical quantities.

We illustrate these below.

To verify the correctness of a physical equation

46

Is the following equation correct, $v = u + at^2$? Where v is the final velocity of an object starting off with an initial velocity, u, accelerating uniformly at a rate, a, after travelling for time, t.

In order to determine the correctness of the equation, we simply check to see if all the terms in the equation have the same dimensions (that is, if the equation is dimensionally homogenous).

There are 3 terms in the equation; v, u, and at^2.

The dimension of v is $\frac{displacement}{Time} = \frac{L}{T} = LT^{-1}$

The dimension of u is same as v (they are both velocities) = LT^{-1}

The dimension of $at^2 = \frac{displacement\ (L)}{Time^2\ (T^2)} \times Time^2\ (T^2) = L$

The 3 terms do not all have the same dimension. Therefore, $v = u + at^2$ is not dimensionally homogeneous, and so the equation is NOT correct.

The equation will have been homogenous if all the 3 terms had the same dimension, LT^{-1}. This could have been achieved if the term (at^2) is modified to (at). You can verify this!

To derive the unit of a physical quantity

Once the dimension of a physical quantity is derived, the unit can be written down.

For example, velocity = $\frac{displacement}{time} = \frac{L}{T} = LT^{-1}$

Therefore the unit of velocity is ms^{-1} (m for length L, and s for time T)

To derive the exact form of the relation between measured physical quantities

48

Suppose the period of oscillation T of simple pendulum depends on the mass m of the pendulum bob, the length l of the thread and the acceleration due to gravity g. We can use the method of dimensions to find the correct relation as follows.

Since T depends on m, l and g, we assume the equation:
$T = k\, m^x L^y g^z$, where k is a dimensionless constant which we may not be able to decide using the knowledge of dimensions alone. x, y and z are arbitral constants which we will determine by using the knowledge of dimensions.

We simply have to ensure that the dimension of quantities on the left hand side is equal to the dimension of quantities on the right hand side.

The dimension of T is T.
For m is M, for l is L, and for g is LT^{-2}.

So we can write:
$$M^o L^o T = k\, M^x L^y (LT^{-2})^z \quad \text{-----------(1)}$$
Notice that we have included $M^o L^o$ to the left hand side just to make sure we have the M and L components. The value of $M^o L^o$ is however 1 and therefore does not change anything about the equation.

Next, we simplify the right hand side to get:
$M^o L^o T = k\, M^x L^y L^z T^{-2z}$
which is same as $M^o L^o T^1 = k\, M^x L^{y+z} T^{-2z}$ -----------(2)

Equating indices of M on both sides of equation (2) gives: $x = 0$ ----(3)

Doing same for L gives: $y + z = 0$ -----------(4)

And for T gives: $-2z = 1$ -----------(5)

Solving equations (3), (4) and (5) gives: $x = 0$, $y = ½$, and $z = -½$.

Substituting these in our original assumption $T = k\, m^x L^y g^z$ gives

$T = k\, m^0\, l^{½}\, g^{-½}$

which can also be written as:

$$T = k\left(\frac{1}{g}\right)^{1/2} \quad \text{or} \quad T = k\sqrt{\frac{l}{g}}$$

Yes, that is how we do it!

Now, You should attempt the following JAMB question

49

Which of the following is the dimension of pressure?
(A) $ML^{-1}T^{-2}$ (B) MLT^{-2} (C) ML^2T^{-3} (D) ML^3T^{-2}

Solution!

50

Pressure $\dfrac{Force}{Area} = \dfrac{Ma}{L^2} = \dfrac{MLT^{-2}}{L^2}$ $ML^{-1}T^{-2}$

So, option (A) is correct

Note!

51

Some quantities are dimensionless!

For example, the quantity strain, defined as $\dfrac{increase\ in\ length}{original\ length}$, is a dimensionless quantity.

In a nutshell!

Table 3. Units and dimensions of physical quantities.

Quantity	Unit	Relation of S.I to base units	Dimension
Speed and velocity	ms^{-1}	ms^{-1}	LT^{-1}
Acceleration	ms^{-2}	ms^{-2}	LT^{-2}
Force	N	$kgms^{-2}$	MLT^{-2}
Volume	m^3	m^3	L^3
Pressure	Pa or Nm^{-2}	$kgm^{-1}s^{-2}$	$ML^{-1}T^{-2}$
Density	kgm^{-3}	kgm^{-3}	ML^{-3}
Work and energy	J	kgm^2s^{-2}	ML^2T^{-2}
Power (work/time)	W	kgm^2s^{-3}	ML^2T^{-3}
Momentum (impulse)	Ns	$kgms^{-1}$	MLT^{-1}
Frequency (cycle/sec)	Hz	s^{-1}	T^{-1}
Temperature	K	K	θ
Specific heat capacity	$Jkg^{-1}K^{-1}$	$m^2s^{-2}K^{-1}$	$L^2T^{-2}\theta^{-1}$
Specific latent heat of fusion or Vapourization	JKg^{-1}	m^2s^{-2}	L^2T^{-2}
Linear expansivity	K^{-1}	K^{-1}	θ^{-1}
Electric current	A	A	I
Electric charge	C	As	TI
Electromotive force or potential difference	V = work/It	$kgm^2s^{-3}A^{-1}$	$ML^2T^{-3}I^{-1}$
Electrical resistance	$R = \dfrac{V}{I}$	$kgm^2s^{-3}A^{-2}$	$ML^2T^{-3}I^{-2}$

And finally! A couple of exercises for you to lay your hands on

Exercises

53

1. The inner diameter of a test tube can be measured accurately using _____
(A) Micrometer screw gauge   (B) Pair of dividers   (C) Meter rule   (D) Pair of Vernier Calipers.   [JAMB]

2. Determine the reading on the vernier scale below.

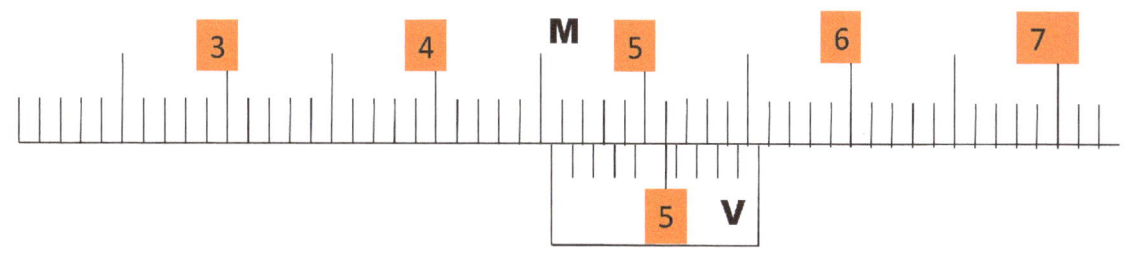

(A) 3.45 cm   (B) 4.55 cm   (C) 4.45 cm   (D) 5.55 cm

3. Determine the reading on the vernier scale below.

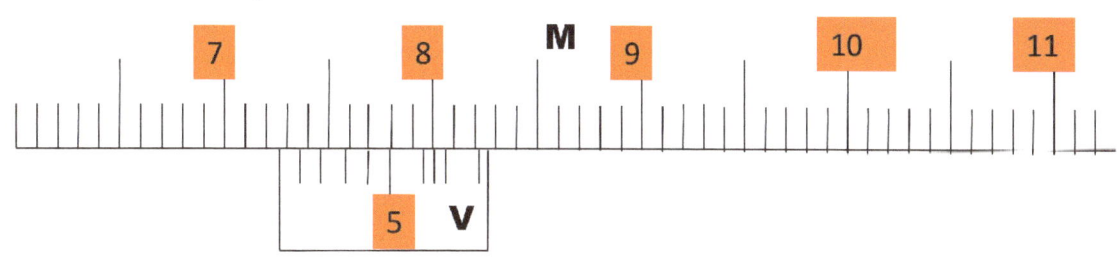

(A) 7.85 cm   (B) 5.78 cm   (C) 8.57 cm   (D) 7.24 cm

4. Determine the reading on the micrometer screw gauge scale below.

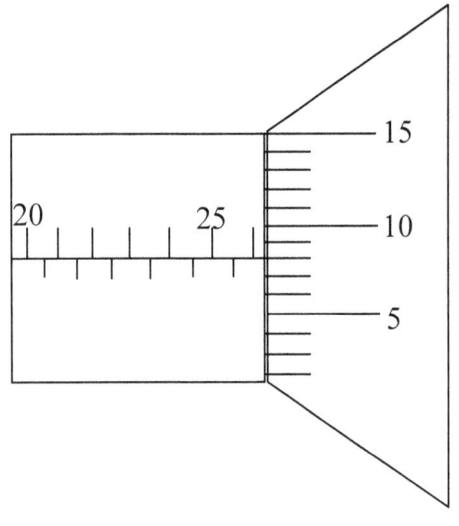

(A) 26.08 mm (B) 26.8 mm (C) 25.10 mm (D) 26.15 mm

5. Determine the reading on the micrometer screw gauge scale below.

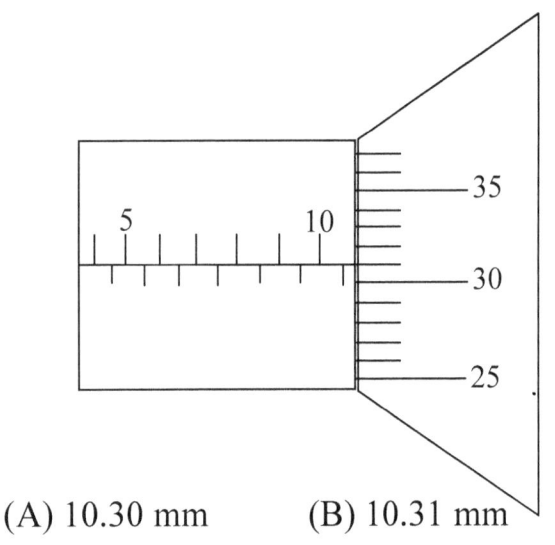

(A) 10.30 mm (B) 10.31 mm (C) 10.81 mm (D) 10.531 mm

6. The following are derived quantities EXCEPT
(A) Velocity (B) Speed (C) Mass (D) Force

7. Impulse is a physical quantity defined as the product of force and time, it is dimensionally equivalent to

(A) Velocity   (B) Pressure   (C) Acceleration   (D) Momentum

8. Given the motion of an object accelerating uniformly at rate, a, from an initial velocity, u, to a final velocity, v, in time, t, and covering the distance, s. The following equations are dimensionally homogenous EXCEPT
(A) $v^2=u^2+2as$ (B) $s=ut+0.5at^2$ (C) $s=0.5(u+v)t$ (D) $v=u+2as$

9. What is the dimension of a physical quantity that is dimensionally equivalent to the ratio $\frac{mass}{acceletation}$

(A) $ML^{-1}T^2$ (B) $ML^{-2}T^2$ (C) MLT^{-2} (D) $ML^{-1}T^{-2}$

10. Which of the following definitions for a physical quantity is dimensionless
(A) $\frac{mass \times energy}{acceletation \times distance}$
(B) $\frac{mass \times acceleration \times distance}{energy}$
(C) $\frac{mass \times acceletation}{distance}$
(D) $\frac{mass \times distance}{acceletation \times energy}$

Solutions

54

1. D
2. B
3. D
4. A
5. C
6. C
7. D
8. D
9. A
10. B